discovermore

Forms of Energy

What Is Light Energy?

Virginia Elliott

Britannica Educational Publishing

IN ASSOCIATION WITH

THE ROSEN PUBLISHING GROUP

Published in 2026 by Britannica Educational Publishing (a trademark of Encyclopædia Britannica, Inc.) in association with The Rosen Publishing Group, Inc.
2544 Clinton Street, Buffalo, NY 14224

Distributed exclusively by Rosen Publishing.
To see additional Britannica Educational Publishing titles, go to rosenpublishing.com.

Editor: Danielle Haynes
Book Design: Michael Flynn

Photo Credits: Cover Piotr Krzeslak/Shutterstock.com; (series background) Dai Yim/Shutterstock.com; p. 4 Iren_Geo/Shutterstock.com; p. 5 Piotr Krzeslak/Shutterstock.com; p. 6 InFocus.ee/Shutterstock.com; p. 7 Top Photo Corporation/Shutterstock.com; p. 8 Dpongvit/Shutterstock.com; p. 9 RugliG/Shutterstock.com; p. 10 Tolkung/Shutterstock.com; p. 11 Dream01/Shutterstock.com; p. 13 (top) Designua/Shutterstock.com; p. 13 (bottom) NigelSpiers/Shutterstock.com; p. 15 (top) Melnikov Dmitriy/Shutterstock.com; p. 15 (bottom) petrroudny43/Shutterstock.com; p. 17 (top) Reshetnikov_art/Shutterstock.com; p. 17 Albert Stephen Julius/Shutterstock.com; p. 18 Frame Stock Footage/Shutterstock.com; p. 19 BlueRingMedia/Shutterstock.com; p. 21 (top) grayjay/Shutterstock.com; p. 21 (bottom) pacfoto/Shutterstock.com; p. 22 LafresiaStockVideo/Shutterstock.com; p. 23 lego 19861111/Shutterstock.com; p. 24 Olga Danylenko/Shutterstock.com; p. 25 Leonid Ikan/Shutterstock.com; p. 26 Duet PandG/Shutterstock.com; p. 27 fuyu liu/Shutterstock.com; p. 29 (top) Jordi Ibern/Shutterstock.com; p. 29 (bottom) Christie Cooper/Shutterstock.com.

Cataloging-in-Publication Data

Names: Elliott, Virginia (Children's writer).
Title: Light energy / Virginia Elliott.
Description: New York : Britannica Educational Publishing, in association with Rosen Educational Services, 2026. | Series: Discover more: forms of energy | Includes glossary and index.
Identifiers: ISBN 9781641905695 (library bound) | ISBN 9781641905688 (pbk) | ISBN 9781641905701 (ebook)
Subjects: LCSH: Light--Juvenile literature.
Classification: LCC QC360.E45 2026 | DDC 535--dc23

Manufactured in China

CPSIA Compliance Information: Batch #QWBRIT26. For further information contact Rosen Publishing at 1-800-237-9932.

Contents

Light and Our World

Light allows people and animals to see. Look around you. Everything you see is **illuminated** by a source of light! Light is a form of energy. Living things need energy to move, work, and grow.

During the day, most of our light comes from the sun. Without the energy from the sun, there would be no plants or animals on Earth's surface. Before humans learned to harness light on their own, nearly all light energy found on Earth came from the sun. Eventually, humans learned to build fires and then to use electricity. This ability to control our environment was a huge step forward for humankind.

Light hits a special part of our eyes called the retina, which then sends signals to our brain that allow us to see.

Energy fuels our universe, from the stars in the sky to the tiniest algae in the sea. Light energy that began its journey out in space allows us to grow our food, light our homes, and do much more. That light then gets transformed into other kinds of energy.

WORD WISE

WHEN THINGS ARE ILLUMINATED, LIGHT IS SHINING ON THEM.

Types of Energy

Light is one of many forms of energy. We can feel heat energy given off by fire. We hear noises thanks to sound energy. All the foods we eat have chemical energy stored in them. Mechanical energy is the energy that a moving object—such as a ball that you kicked—has because of its movement and position. Electrical energy is associated with the flow of tiny particles called electrons between atoms. Atoms are the tiny units that make up everything in the universe.

A fire gives off heat energy, allowing us to roast marshmallows. Those marshmallows have chemical energy, which we eat to fuel our bodies.

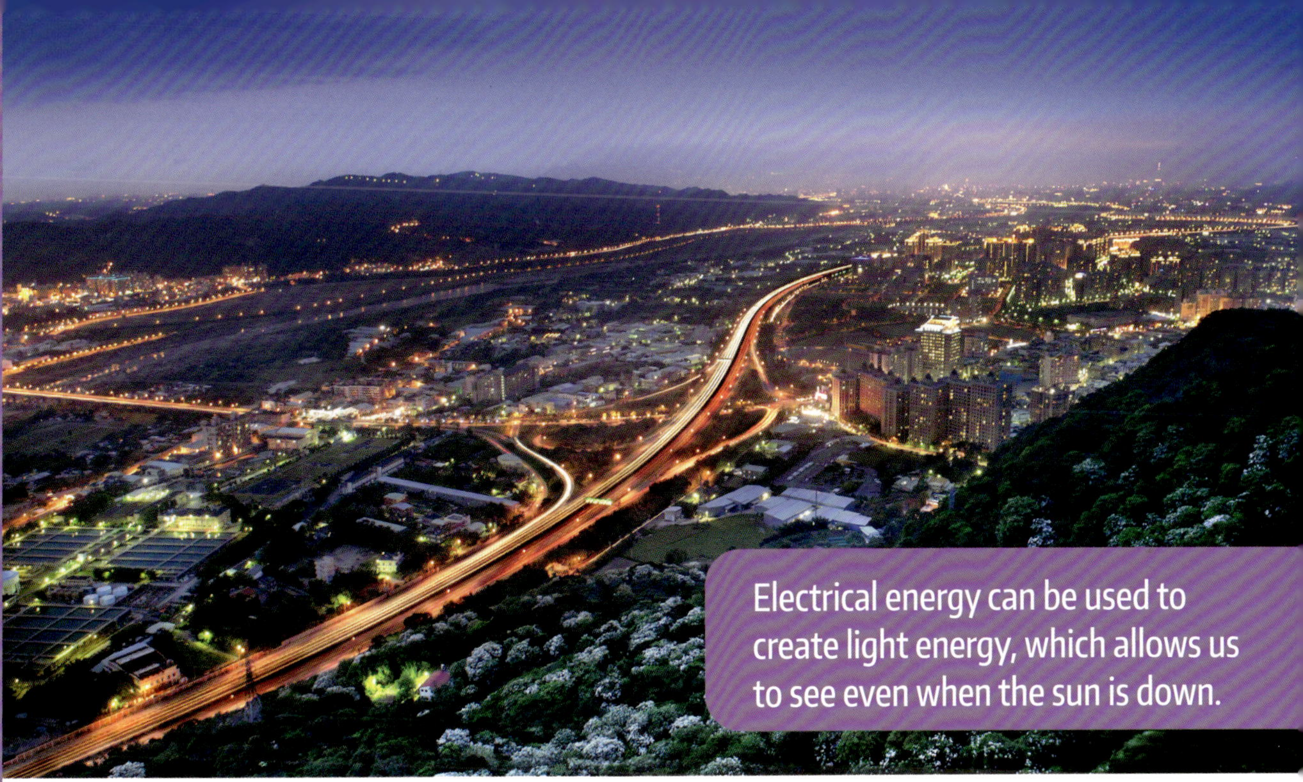

Electrical energy can be used to create light energy, which allows us to see even when the sun is down.

Energy can be transformed, or changed, from one form to another. Lightbulbs let us transform electrical energy to light energy. Chemical energy from burning gasoline transforms into mechanical energy that powers our cars and trucks. The electrical energy from a lightning bolt can strike the ground and give off heat energy. Energy is never completely destroyed; it just changes into another form.

Consider This

What are some examples of different forms of energy that you have noticed around you?

Where the Light Comes From

The sun isn't the only source of light in our world. Fire and burning objects give off light and heat. For example, torches, candles, and oil lamps are all sources of light. When they burn, a chemical reaction occurs that releases the energy we see as light. These were humans' first sources of light after sunlight.

Lightning is another natural source of light. Lightning is a form of electricity that is released from Earth's atmosphere during thunderstorms. Humans learned to produce electrical energy using chemical, mechanical, and **nuclear energy**. Now electrical energy is used to power lightbulbs.

An oil lamp creates light and heat energy by transforming the chemical energy in oil, a type of fuel.

Some very small ocean animals create a source of light called bioluminescence.

Some animals, such as fireflies and certain fish, can give off light. Plants and animals that glow in the dark—glowworms, fireflies, and some mushrooms—change the chemical energy stored in their tissues to light energy. Organisms might use this light to attract a mate or food, or to defend against predators.

WORD WISE

NUCLEAR ENERGY IS THE ENERGY THAT HOLDS TOGETHER THE NUCLEUS, OR CENTER, OF AN ATOM. PULLING APART A NUCLEUS, OR SQUISHING TOGETHER TWO NUCLEI, PRODUCES A LOT OF ENERGY.

Wavelengths

The physical existence of light is a little complicated. Light exists in two different forms at the same time. One form is tiny particles that scientists believe have no mass called photons. The other form of light is waves. The easiest way to think about light is as waves, like those in the ocean. Light waves travel through space at extremely fast speeds.

Light travels 186,000 miles (300,000 kilometers) per second. Still, the light from the closest star to our solar system, Alpha Centauri, takes 4.3 years to travel to Earth!

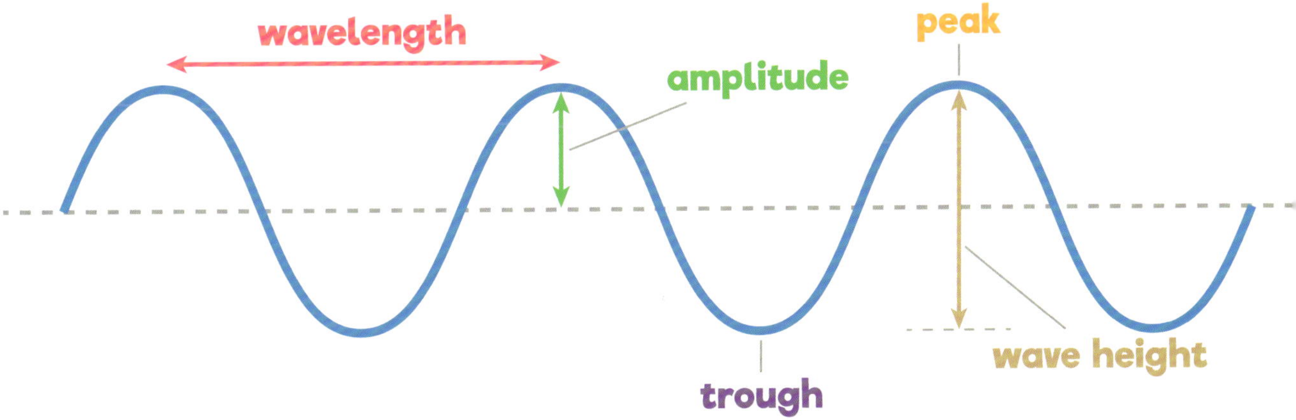

Wavelength is the distance between two peaks or troughs, and amplitude deals with the height of a wave.

Scientists use an idea called wavelength to describe light waves. Like waves moving across a pool of water, light waves have peaks and valleys (or troughs). The distance between two of these high or low points is called a wavelength. The sunlight we see each day contains light waves with long, medium, and short wavelengths. We see light of some wavelengths, but not all, with our eyes. To see other wavelengths, we need special tools such as night-vision goggles or infrared cameras.

Consider This

Can you imagine how fast light travels? Light waves travel through space at a speed of about 186,282 miles (299,792 kilometers) per second.

Types of Radiation

Radiation is the movement of energy in the form of waves or particles. Light, sound, and heat are all examples of radiation. The two main categories of radiation are electromagnetic and mechanical.

Light is electromagnetic radiation. Electromagnetic radiation comes from atoms, the building blocks of all matter. Some of the particles that make up atoms have an electric charge. The motion of these charged particles produces energy that can be described as traveling as waves. These waves can travel through empty space, like in outer space. They can also travel through air and even solid substances, such as glass.

Mechanical radiation cannot travel through empty space. It can only travel through a substance, such as air, water, or a solid object. An example of mechanical radiation is the shock waves created by an earthquake, which travel through Earth and cause the ground to shake. Another example of mechanical radiation is sound energy. Sound waves travel more slowly than light.

Some radiation is bad for humans. Ultraviolet (UV) radiation from the sun can harm our skin, but the Earth's ozone layer blocks out some of these rays.

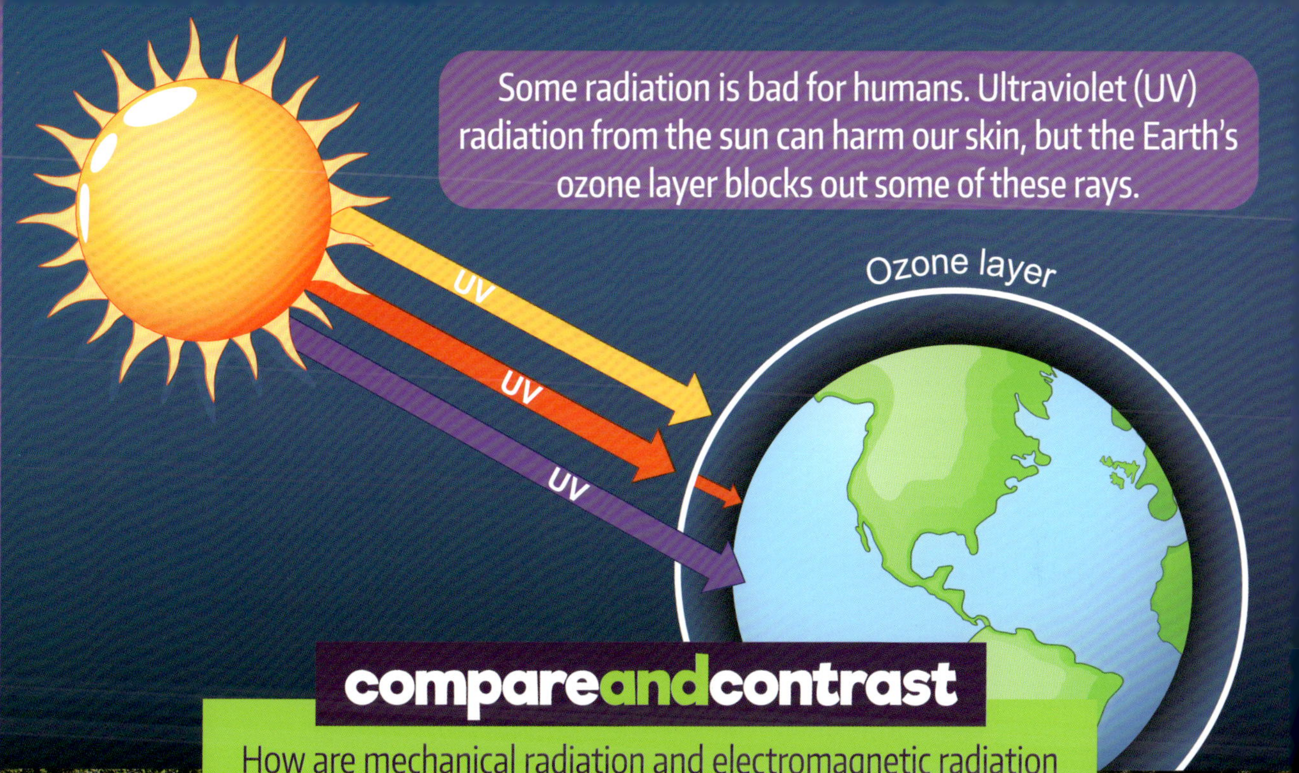

compareandcontrast

How are mechanical radiation and electromagnetic radiation similar? How are they different?

The mechanical radiation of an earthquake causes movement in the ground, which can break and cause damage.

Seeing Colors

We need light to be able to see color. When people see colors, they are really seeing different types of light bouncing off objects.

The color of light depends on its wavelength. Each color has its own wavelength. Red light has the longest wavelength. Violet light has the shortest wavelength. Orange, yellow, green, blue, and indigo have wavelengths in between those of red and violet. When light waves of all the colors travel together, the light looks white. The range of colors is called the **visible spectrum**. It consists of all the colors we see around us. The visible spectrum is a small part of the total range, or spectrum, of all electromagnetic radiation. Other kinds of radiation, including X-rays and radio waves, have longer or shorter wavelengths than those of visible light. Infrared light, however, has a longer wavelength than visible light.

WORD WISE

THE VISIBLE SPECTRUM CONTAINS ALL THE COLORS OF VISIBLE LIGHT, WHICH INCLUDES RED, ORANGE, YELLOW, GREEN, BLUE, INDIGO, AND VIOLET LIGHT.

Remote controls use infrared light to change channels, but you can't see it normally.

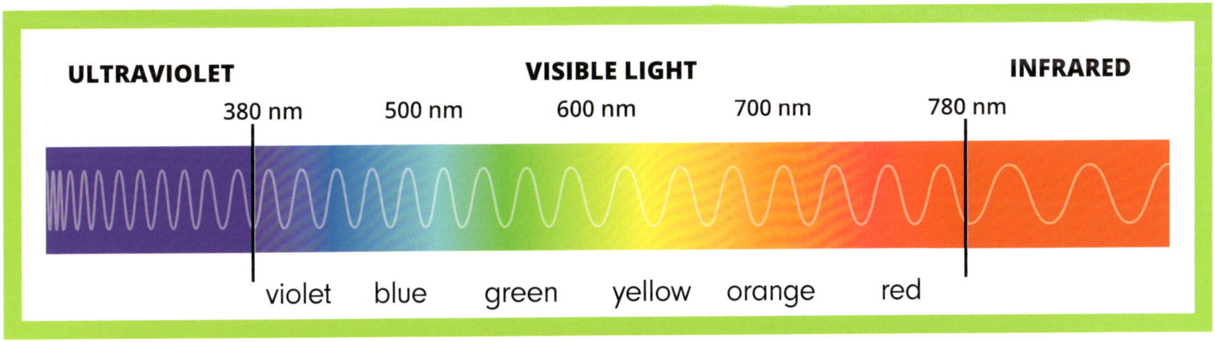

We can't see light with the shortest or longest wavelengths. Wavelengths that are between 380 nanometers (nm) and 780 nanometers, though, can be seen by the naked eye.

Rainbows

Have you ever noticed what looks like a rainbow of light on the wall or floor inside your home? When white light passes through a special piece of glass called a prism, the light bends. Waves of different wavelengths bend by different amounts. The longest wavelengths bend the least. The shortest wavelengths bend the most. This bending separates the light waves so each color can be seen.

Prisms are found in binoculars, microscopes, and other instruments. Scientific instruments called spectroscopes use special prisms to separate light into a very wide spectrum of colors. This allows scientists to study the light from distant stars and other bodies in space.

Drops of water can also act as prisms. When light waves pass through raindrops, the light waves separate. This happens because the raindrops bend the light waves just as a prism does. The separated light waves appear as a rainbow, or a multicolored arc, in the sky. You have a good chance of seeing a rainbow when it's sunny just after a rainstorm.

Seeing "rainbows" indoors might seem a little strange. Don't they only happen outdoors?

compareandcontrast

How are raindrops like prisms? How are they different?

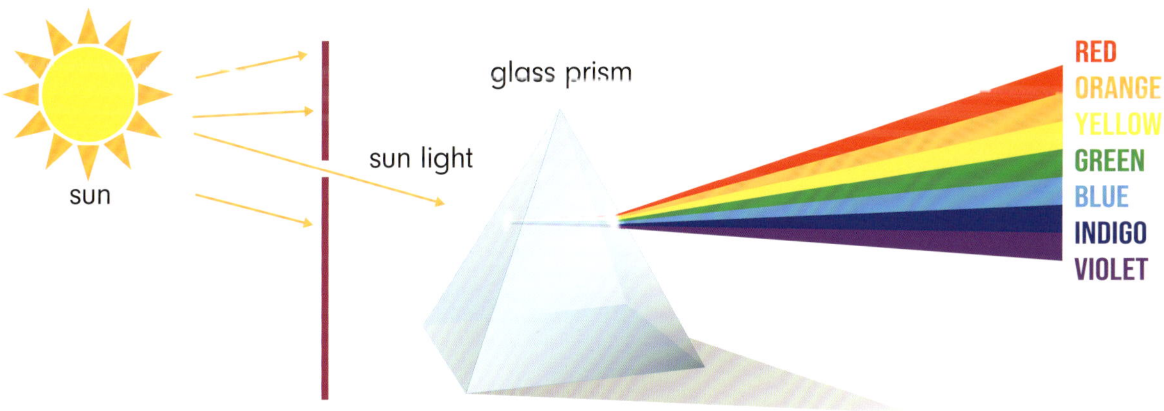

Rainbows indoors and outdoors are made the same way—light passes through a glass prism (indoors) or raindrops (outdoors) and creates a spectrum of colors.

Reflection and Sight

An object can be seen only if light travels from the object to an eye that can sense it. Light can travel to an eye when an object gives off light or when an object reflects, or bounces back, light.

People see the sun because it gives off its own light. But people see the moon because light from the sun reflects off of it.

The moon reflects sunlight back to Earth even after the sun has gone down.

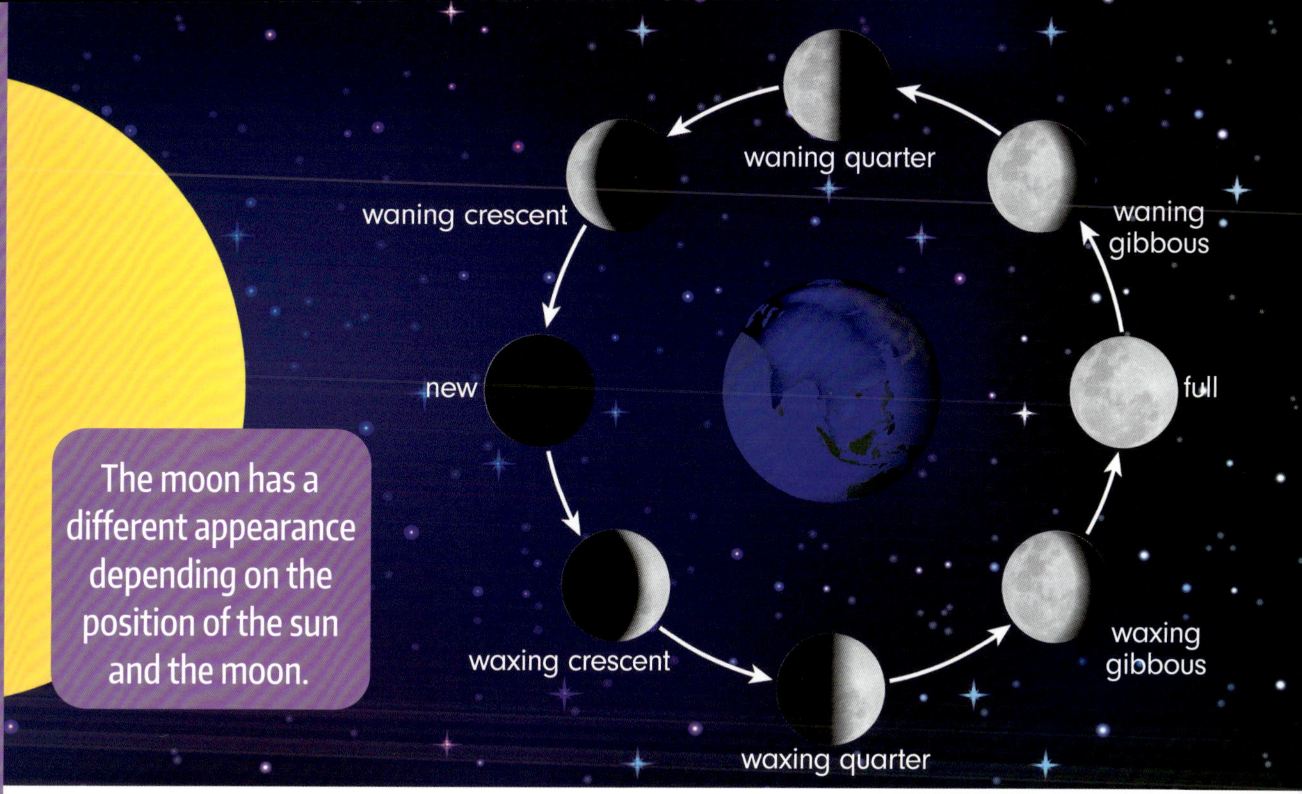

The moon has a different appearance depending on the position of the sun and the moon.

At night, the moon is illuminated by the sun. Although we do not see the sun at night, it is still shining from behind our planet. The moon and Earth are constantly moving, so light from the sun hits the moon and bounces toward Earth in different ways every day. As a result, a person standing on Earth sees different views of the moon at different times. These varying appearances are called phases. The moon repeats all its phases every 29.5 days.

Consider This

Why do we see stars in the night sky?

Opaque or Transparent?

Why can you sometimes see through an object? When only the surface of an object (such as the moon) can be seen, the object is said to be opaque. Opaque objects do not let light pass through them. Instead, they reflect and absorb light. As a result, they are not see-through. A mirror is an example of an opaque object that reflects almost all of the light that hits it.

A colored object reflects only certain wavelengths of light. The reflected light is the color the object appears to be. For example, a red apple absorbs all the colors of light that hit it except for red light.

Materials that do not absorb or reflect much light are transparent, or see-through. Clear glass is transparent because most of the light that hits it passes through it. Objects that allow some, but not all, light to pass through them are called translucent. Stained-glass or frosted windows are translucent.

We see some apples as red because that's the only color of light they reflect. Green apples reflect green light.

compareandcontrast

How do opaque and transparent objects differ? How are they similar?

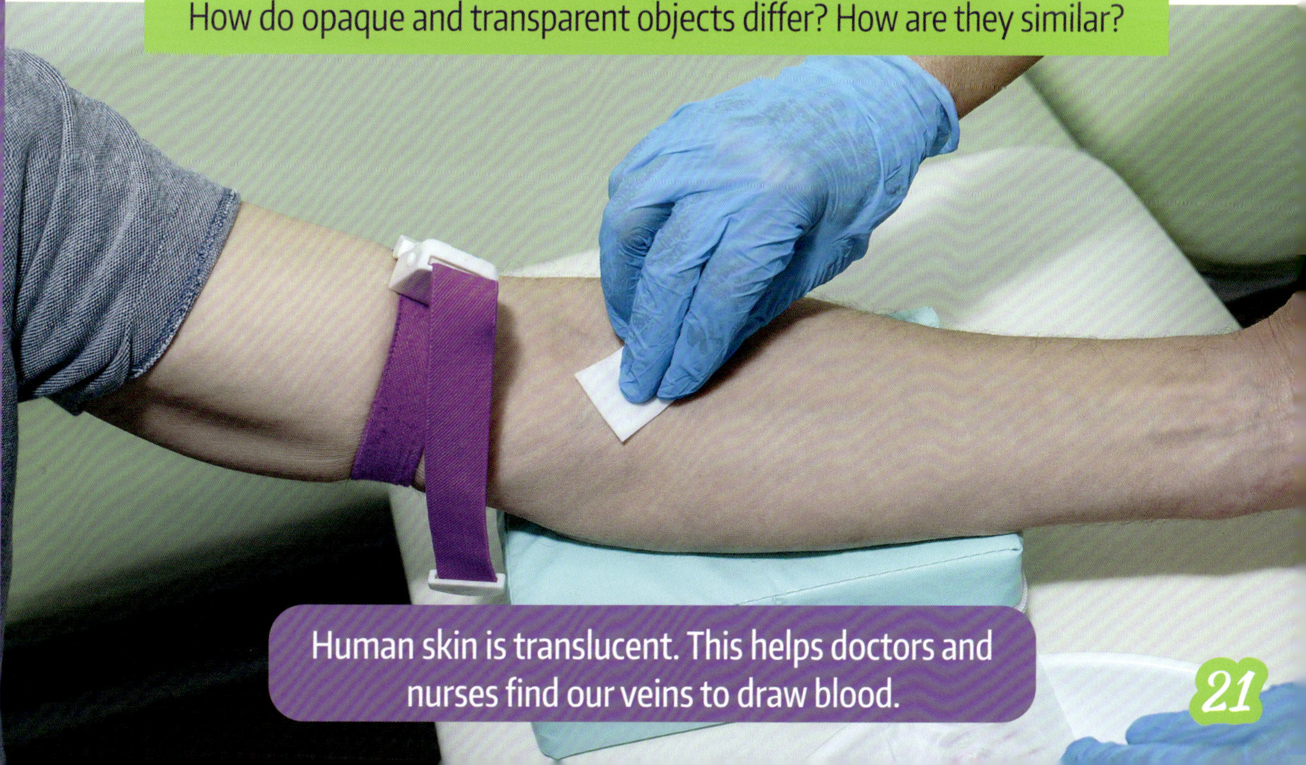

Human skin is translucent. This helps doctors and nurses find our veins to draw blood.

How Animals See

Most animals—including humans—depend on sight to do many things. During the day, sunlight lights the world and makes it possible for humans to see. At night, we are unable to see unless we produce our own light.

Humans see light in what is called the visible range, which includes all colors of light seen in a rainbow. Other animals see or sense different kinds of light.

The pit organs on a pit viper are little indentations found between the snake's eye and nostril.

Bees have five eyes—two large ones that have many lenses and three small ones that help them see UV light.

Snakes called pit vipers, for example, have sense organs (called pits) that "see" rays that humans feel as heat. These rays are called infrared radiation. These pits help them sense heat coming from the body of their prey. Once a pit viper senses the heat, it can attack its prey even in the dark. Bees see some of the colors that humans see, but they are also sensitive to ultraviolet radiation. Ultraviolet (UV) radiation is beyond the range visible to most humans. People born without a lens in their eyes or who have had them removed, might be able to see UV light!

Consider This

How might flowers in a garden appear to a pit viper or a bee?

Plants and Life

Life on Earth doesn't exist without light energy. Without it, there would be no green plants, and without plants there would be no animals. Through a process called photosynthesis, plants use sunlight to make their own food. Photosynthesis requires sunlight, chlorophyll, water, and **carbon dioxide** (CO_2). Chlorophyll is a substance in all plants, especially in the leaves. Plants take in water from the soil and carbon dioxide from the air.

Rainforests are especially important to humans, because they help produce the oxygen we breathe.

During photosynthesis, plants take light energy from the sun, carbon dioxide from the air, and minerals from the ground to create chemical energy in the form of sugars and oxygen.

Photosynthesis starts when chlorophyll absorbs energy from sunlight. Green plants use this light energy to change water and carbon dioxide into oxygen and nutrients called sugars. The plants use some of the sugars and store the rest. The oxygen is released into the air.

People and other animals need oxygen to breathe to stay alive. They also rely on plants for food. They gain energy from the chemical energy that is stored in the sugars made during photosynthesis. Another word for these sugars is glucose.

WORD WISE

CARBON DIOXIDE IS THE GAS FORMED WHEN PEOPLE AND OTHER ANIMALS BREATHE OUT. IT IS AN IMPORTANT PART OF EARTH'S ATMOSPHERE.

Solar Energy

Solar energy is what we call all the energy that comes from the sun, including light and heat. It fuels both plants and animals. Solar energy is the source of light for photosynthesis. It is also the source of heat that plants and animals need to survive. The heat from the sun causes water on Earth's surface to evaporate and form clouds that eventually provide fresh rainwater.

You might notice cats love to take naps in sunbeams. That's because it's probably the warmest spot in the house!

Solar panels can be installed on top of buildings, or hundreds of them can be placed out in open fields to collect sunlight.

Energy from the sun can also be collected and used to heat buildings and to make electricity. Most solar heating systems capture solar energy with a device called a flat-plate collector. The collector is a large plate of black metal covered with a sheet of glass. It is usually placed on the roof of a building. The plate absorbs sunlight and uses it to heat air or water that flows through pipes behind it. We call solar energy a renewable source of energy because we never run out of it.

Consider This

What different uses might there be for solar energy?

Renewable Energy

There are other ways to use solar energy. A solar furnace uses the sun's heat by focusing light with mirrors to heat a small area. The heat is used to make steam. The steam can be used to make electricity in a power plant. Solar cells use the sun's light, turning it into electricity. A single solar cell makes only a little electricity, but groups of solar cells can provide electricity for whole buildings. Large groups of solar cells are called solar panels. Solar cells are also used in such products as calculators and watches.

Solar energy has two big benefits over fossil fuels like coal, oil, and natural gas. First, fossil fuels can be used up, but there is an endless supply of sunlight. Second, solar energy does not cause pollution, like burning fossil fuels does.

The world's largest solar furnace is in France. The Odeillo solar furnace is a research site where scientists study very high temperatures.

compareandcontrast

How are fossil fuels and solar energy different? How are they the same?

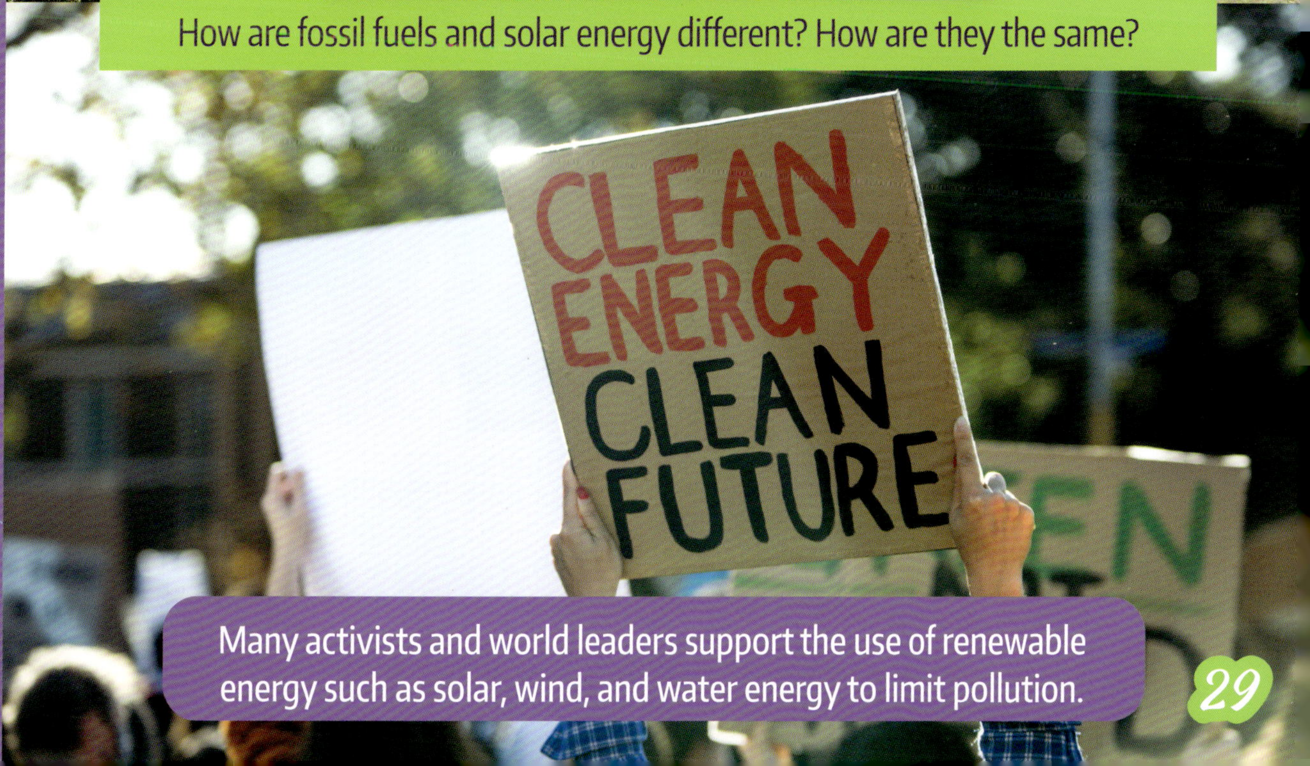

Many activists and world leaders support the use of renewable energy such as solar, wind, and water energy to limit pollution.

Glossary

chemical reaction A chemical change that occurs when one or more substances combine to form a new substance.

electromagnetic radiation Energy in the form of electromagnetic waves, such as visible light, X-rays, and radio waves, that comes from the motion of charged particles in atoms.

electron A very small elementary particle within an atom that has a negative charge of electricity and travels around the nucleus of an atom.

evaporate To change into a gas.

fossil fuel A fuel, such as coal, oil, or natural gas, formed in the ground from plant or animal remains.

infrared Relating to rays like light but lying outside the visible spectrum at its red end.

nucleus The center of an atom.

oxygen One of the main elements that make up air and that is necessary for the survival of all plants and animals.

particle One of the very small parts of matter (as a molecule, atom, or electron).

photon A tiny bit or bundle of electromagnetic radiation.

prism A transparent object that usually has three sides and bends light so that it breaks up into rainbow colors.

solar cell A photoelectric cell that converts sunlight into electrical energy and is used as a power source.

ultraviolet radiation A kind of electromagnetic radiation having a wavelength shorter than that of visible light but longer than that of X-rays.

wavelength The distance from the peak of one wave to the peak of another one.

X-ray A kind of electromagnetic radiation of a very short wavelength that can move through objects and act on photographic film as light does.

For More Information

Books

Bunting, Philip. *Super Power: Renewable Energy: What It Is, How We Get It, and Why We Need It.* Richmond, Victoria, Canada: Bright Light, 2024.

Hughes, Susan. *Lights Day and Night: The Science of How Light Works.* Toronto, Canada: Kids Can Press, 2021.

Lundgren, Julie K. *Light: Energy We Can See!* New York, NY: Crabtree Publishing Company, 2022.

Websites

Britannica Kids: Light
www.kids.britannica.com/kids/article/light/353386
Learn more about light energy with more diagrams and related articles.

Ducksters: Science of Light
www.ducksters.com/science/light.php
Take a fun quiz on this website about light energy.

Index